# Math and Science Games
## for Leadership

# Math and Science Games
# for Leadership

**Seah Wee Khee**
**Sukandar Hadinoto**
**Charles Png**
**Ang Ying Zhen**
**2nd Student Council**

NUS High School of Mathematics and Science, Singapore

NUS
HIGH
SCHOOL
of Math & Science

**World Scientific**

NEW JERSEY · LONDON · SINGAPORE · BEIJING · SHANGHAI · HONG KONG · TAIPEI · CHENNAI

*Published by*

World Scientific Publishing Co. Pte. Ltd.

5 Toh Tuck Link, Singapore 596224

*USA office:* 27 Warren Street, Suite 401-402, Hackensack, NJ 07601

*UK office:* 57 Shelton Street, Covent Garden, London WC2H 9HE

**Library of Congress Cataloging-in-Publication Data**
50 math & science games for leadership / by Seah Wee Khee ... [et al.].
    p. cm.
  Fifty math & science games for leadership
 Fifty math and science games for leadership
 ISBN-13: 978-981-270-692-8
 ISBN-10: 981-270-692-5
 1. Leadership--Study and teaching (Secondary)--Activity programs. 2. Mathematics--Study and teaching (Secondary)--Activity programs. 3. Science--Study and teaching (Secondary)--Activity programs. I. Khee, Seah Wee.

HM1261.A15 2007
303.3'40712--dc22

2007005961

**British Library Cataloguing-in-Publication Data**
A catalogue record for this book is available from the British Library.

ISBN-13 978-981-270-692-8
ISBN-10 981-270-692-5

Inhouse editor: Juliet Lee Ley Chin

Printed in Singapore by Mainland Press

# Acknowledgements

We would like to thank several people whose encouragement has made it possible to complete this book well.

- ❖ Our school principals, Associate Professor Lai Yee Hing, Mr Suresh Balakrishnan and Mrs Jennifer Phang. Special thanks to Mrs Phang for her support and encouragement to express ourselves creatively.
- ❖ Dr K K Phua, for his contribution and support for making the publication of this book possible for us.
- ❖ Dr Vivian Balakrishnan, Minister of Community Development, Youth and Sports, whose ways and words have inspired us to develop this resource book for our community.
- ❖ Our loved ones, including our spouses and parents for their kind understanding and patience with us.
- ❖ The 2nd Student Council of NUS High School for thinking through the games with us and giving us invaluable feedback.
- ❖ Mr Samuel Lee for his games, Mr Syed Mahdar for his art work, Andre Tan for his cartoons, and the Affective & Character Education team at NUS High School, for their contribution of games and pictures.
- ❖ Our colleagues in the Biology, Chemistry and Physics Departments for their kind understanding and patience as we did our writing.
- ❖ Our personal mentors, whose words of support meant many things to all of us.

# CONTENTS

# PART II SCIENCE GAMES

# Preface

We believe teaching leadership through games is an effective way to get the message of leadership across. But this book is not just a collection of leadership games. It is a work of love to integrate the elements of Math and Science in teaching the leadership principles.

Each game exercise includes a game activity and a review question component comprising "Process" and "Practical Application"—offered as the last two sections of each game exercise—which will stimulate discussion and reflection on the key leadership concepts.

The Matrixes in Figure 1 and Figure 2 will provide you with a quick glance of the leadership concepts and the games that will elicit the key leadership understanding emphasised in this book. Figure 1 comprises the first 24 game exercises highlighting mathematical concepts or principles, while Figure 2 comprises the last 26 games highlighting scientific concepts or principles.

### How each leadership game is organised

Each of the 50 games has the following ten components, except for "Possible Variation(s)" which is provided additionally in some of the games.

*Key Leadership Understanding*   Highlights the key purpose of the activity. It identifies the learning outcomes in terms of what the players should be able to articulate at the end of the activity

*Math/Science Concepts Applicable*   Identifies the key Math or Science concepts or principles that are tested, so players who are more familiar with such concepts or principles might do better in the relevant games.

*Equipment/Logistics*   Lists the equipment and resources required for conducting the game

*Time Required*   Indicates an estimated amount of time needed to carry out the game activity, excluding the review question component comprising "Process" and "Practical Application"

*Game Objective*   Refers to the objective or requirement of the game—what each group of players has to do to be the winner from the game activity *per se*, excluding the review question component

*Group Size*   Indicates the minimum, maximum or an estimated ideal number of players for each team

*Procedure*   Explains the step-by-step instructions and actions to be carried out by the facilitator with the players when conducting the game

*Possible variations (where applicable)*   Suggests various ways to modify a particular game

*Process*   Lists questions for players to reflect on what went on during the game activity itself (first part of review question component)

*Practical Application*   Lists questions for players to relate what they have reflected from the game activity to what is happening or could happen in their own team or organisation outside of the game activity (second part of review question component)

## What is the best way to use this book?

Decide on the focus of your leadership lesson. What values and traits do you wish to highlight to your group of trainees? At the same time, decide on whether you would prefer a game with a tilt towards mathematics or science.

At the right-hand top of each Game, you will see an oval-shaped header indicating the level of difficulty of the game—"Elementary", "Intermediate" or "Advanced". This refers to the game activity *per se*, and not necessarily the whole game exercise itself which would include the review question component. How "elementary" or "difficult" the review question component is depends on how you conduct the discussion, the length of time given to participants to share and articulate their feedback and reflections, the sensitivity experienced by the participants, the camaraderie or comfort level among the participants, and so on and so forth.

## In conclusion

Similar to many resource guides, the game exercises in this collection are meant as an aid. Each exercise can be modified to suit the user's needs. Game 18, for example, requiring players to know multiples of 7, can be modified to require players to know multiples of numbers smaller or bigger than 7. The review question component in each game exercise can also be omitted, if you are keen on just the game activity without a leadership training agenda in mind for your players.

Suffice to say, every game activity takes on a shape or shade of its own, when played.

Be prepared.

Be surprised.

Have fun!

Fig. 1. Leadership Game Matrix (for Math).

| Leadership Understandings/ Games | Ask Questions | Create & See Things Differently | Develop Resources | Be Disciplined | Active Listening | Make Priorities | Setting a Good Example for Others to Follow | Multiply Leaders | Problem Solve | Sacrifice | Search and Explore | Strategize | Support Diversity | Team work & Collaborate |
|---|---|---|---|---|---|---|---|---|---|---|---|---|---|---|
| Addition | X | | | | | | | | X | | | X | | |
| Guess The Number | | | | X | X | | | | X | | | | X | X |
| "Beengo!" | | | | | X | | | | X | | | X | X | X |
| Symmetry | | X | | | | | X | | | | X | | | X |
| Multi-Division | | | | X | X | | | X | | X | | X | | X |
| Another PIE Jam | X | | | | X | | | | X | | | X | X | X |
| Blindfold Polygons | X | X | | | X | | | | X | | | X | X | X |
| Money Mind | | | X | | | X | | | X | X | | X | | X |
| 7 up with a Twist! | | | | X | X | | | | | | | | X | |
| Shape Me! | | X | | | | | | | X | | | X | X | X |
| Momentum | | | | X | X | | X | | | | | | X | X |
| Graphs | X | | X | | X | | | | | | | X | X | X |
| Aladdin's Magic Carpet | | X | X | X | X | | | | X | | | X | X | X |
| Line Up! | | | | X | X | | | | | | | X | | X |
| Fishing | | | X | | | | | | X | | | X | | X |

Fig. 1. (*Continued*)

| Leadership Understandings/ Games | Ask Questions | Create & See Things Differently | Develop Resources | Be Disciplined | Active Listening | Make Priorities | Setting a Good Example for Others to Follow | Multiply Leaders | Problem Solve | Sacrifice | Search and Explore | Strategize | Support Diversity | Team work & Collaborate |
|---|---|---|---|---|---|---|---|---|---|---|---|---|---|---|
| The Olympics of Science and Math | | | | | | | X | | | | | X | | |
| Angle of Elevation | | | | | | | | | | | | X | X | X |
| In Step with the Times! | | | | | | | | | X | | | X | | X |
| Whose Bubble Is It? | | | X | | | | | | | | | | X | |
| Number Game | | X | | X | X | | | | X | | | X | | X |
| Vector Pull | | X | X | X | X | | | | X | | | X | X | X |
| Danger Zone! | | | | | X | | | | | | X | X | | X |
| All on a Square | | | | | | | X | | | X | | | | X |
| Station Omega | | X | | X | X | | | | X | | | X | | X |

Fig. 2. Leadership Game Matrix (for Science).

| Leadership Understandings/ Games | Ask Questions | Create & See Things Differently | Develop Resources | Be Disciplined | Active Listening | Make Priorities | Setting a Good Example for Others to Follow | Multiply Leaders | Problem Solve | Sacrifices | Search and Explore | Strategize | Support Diversity | Team work & Collaborate |
|---|---|---|---|---|---|---|---|---|---|---|---|---|---|---|
| Rubber Banding | | | | | | | | | | | | | | X |
| The World's Your Oyster | | X | X | | | | | X | | | X | | | X |
| Once upon a Time | | X | X | | X | | | | | | X | | | X |
| Drinks for Life | | X | | | | | | | | | X | | | X |
| Biodiversity | | | | | X | | | | | | | | X | |
| Charades | X | X | | | | | | | X | | | | | X |
| Table of Elements | | X | | | | X | | | X | | | | | |
| Nervous Pulses | | | | | | | | | | | | | X | X |
| Group Obstacle Race | | X | X | X | | X | | X | X | | X | X | X | X |
| Primordial Soup | | | | | X | | | X | X | X | | X | X | X |
| Height Equilibrium | X | X | X | | | | X | X | X | | X | X | X | X |
| Treasure Hunt | X | | X | X | X | X | | X | X | X | X | X | X | X |
| Scientific Scrabble! | | X | | | | | | | X | | X | | | X |

Fig. 2. (Continued)

| Leadership Understandings/ Games | Ask Questions | Create & See Things Differently | Develop Resources | Be Disciplined | Active Listening | Make Priorities | Setting a Good Example for Others to Follow | Multiply Leaders | Problem Solve | Sacrifices | Search and Explore | Strategize | Support Diversity | Team work & Collaborate |
|---|---|---|---|---|---|---|---|---|---|---|---|---|---|---|
| Murder Mystery | X | | | | X | | | X | X | | X | | | X |
| Pull It Up! | | X | X | | X | | | X | X | | X | X | | X |
| Food Web | | | | | | | | | X | X | | | | |
| Cam the Chemicals! | | X | | | | X | | X | | | X | | | X |
| Taboo! | X | | | X | X | | | | X | | | | | X |
| Circle of Influence | | | | X | | | | | | | | | X | |
| Boeing or Airbus? | | X | X | | | | | | X | | X | X | | X |
| Bob the Builder | | X | X | X | X | | | X | X | X | | X | X | X |
| Water Bomb Volleyball | | | | X | | | | | | | | X | X | X |
| Swoosh! | | | | X | | | | | | X | | X | | X |
| Mahjong Dominoes! | | X | X | | X | | | | | | X | | | X |
| Leucocytes | X | | | X | X | | | | | | | | | X |
| Broken Telephone Line/Fax Machine | | | | | X | | X | | | | | | | |

# About the Authors

## SEAH Wee Khee

Wee Khee is an Education Officer in Biology and Student Council teacher-in-charge, focusing on Leadership Development at the NUS High School of Mathematics and Science. Graduated with a PhD from the National University of Singapore (NUS) in Life Sciences, she is currently pursuing a Masters of Educational Management (MEM) at the University of Melbourne, Australia. She was awarded a NUS Research Scholarship and was the recipient of the prestigious President Graduate Fellowship. To date, she has also published several research papers in internationally reviewed journals.

## Sukandar HADINOTO

Sukandar is currently a Chemistry teacher at the NUS High School of Math and Science. He graduated with a BSc Hons (NUS). A Student Council teacher and an NCC teacher, Sukandar enjoys being with young people. He also likes adventure training stints and loves the outdoors.

## Charles PNG Soon Hock

Soon Hock is currently Head of Student Development at NUS High School of Mathematics and Science where he focuses on Affective & Character Education, Service Learning and Leadership Development. Soon Hock completed his scholarship bond with the Public Service Commission and holds an MBA (AGSB) and a BA Hons (NUS). He is a trained facilitator in the use of Kouzes and Posner's Leadership Challenge Model and also a trained facilitator in the 6 Seconds model of Emotional Intelligence. His experience in inspiring young individuals includes mentoring player groups to win Citibank-YMCA Youth for Causes awards.

## ANG Ying Zhen

Ying Zhen is a Year 5 student who is also the President of the Player Council at the NUS High School of Mathematics and Science. She continues her secret love for basketball and debates in her free time in school.

## 2<sup>nd</sup> Student Council

The NUS High School Student Council is the premier leadership body in NUS High School, consisting of 30 members—ranging from Year 2s to Year 5s (ages 14 to 17). A young council (only in their 2<sup>nd</sup> year), they strive towards their vision: *Ambassadors of the School, Servants of Community*, and work hand in hand with the school management and the student body to make school life more enjoyable for everyone. They are Margaret AGUSTINA, ANG Moh Lik Roy, ANG Ying Zhen, CHEN Ting An, CHUA Kai Ting Cheryl, FUNG Ai Wei, Dorothy Hannah HUANG Min, JIANG Yu Heng, KOH Zhiwen Sidwyn, KOK Xiu Ling Florence, KWAN En Qi Angela, LAI Chui Yi, LAU Kang Ruey Gregory, LEE Cheng Feng Gary, LEE Eun Kyung, LEE Yun Zhi, LEE Zi-en Bryan, LIM Mingjie Kenneth, NG Wen Bin Reico Maynard, ONG Tien Sheng Royston, POH Ee Leng, SOH Wei Zhi Andy, SYN Mao Ke Mikel, TAN Wan Yu, TAN Zhong Ming, Viona LAM Xin Yi, WONG Shin Nee Samantha, XIONG Qian Cheng and ZHAO Ye.

# PART I
# MATH GAMES

**Game 1: Addition**

1. Each group is given 4 dice.

2. Group members roll the dice to generate numbers.

a 6!  I have a 1!

3. Either add, subtract, multiply, or divide those numbers to try and reach exactly 100.

6+1=7.
7×6=42...

4. First group to reach 100 wins.

101-1=100!

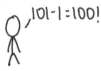

# Leadership Game 1: Addition

*Key Leadership Understanding*

Leaders depend on others. Interdependence and teamwork are crucial elements in leadership.

*Math/Science Concepts Applicable*

Addition

*Equipment/Logistics*

Four dice for every group of participants
Group list
Recorder of group scores

*Time Required*

As long as it takes for the first group to reach a total of 100

*Game Objective*

Be the group to find the fastest way to reach 100 with a pair of dice each

*Group Size*

Any number about two (but preferably less than 10).

*Procedure*

Group participants in teams of two to four. Each participant takes turns tossing the dice and adding the total of both dice. The score the participant gets is the total for that round. Each score is added to the earlier score. The game is over when a team reaches a predetermined score of 100.

## Possible Variations

Why restrict the players to just addition? You may get the players to make use of multiplication, division, and subtraction with the values they generate with the dice. Get them to think carefully with their values before they calculate. Alternatively, the predetermined score can be altered.

## Process

- What difficulties did you face as a group?
- What feelings did you experience when you failed to reach 100/exceeded 100?
- Who was the leader among the group? Would having a leader help in the process?

## Practical Application

- What do you think are the important elements of teamwork?
- What are we doing to build team unity and confidence?
- What takes us so long to recognize the true extent of our problems?

## Game 2: Guess the Number

1. Facilitator thinks of number.

2. Group members ask yes/no questions.

3. Facilitator answers, group members continue asking questions.

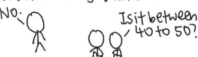

4. Group guesses the number.

# Leadership Game 2: Guess the Number

### Key Leadership Understanding

Leaders listen and ask pertinent questions. The art of listening and asking questions is important in being an effective leader.

### Math/Science Concepts Applicable

Approximation and patterns

### Equipment/Logistics

List of participants
Recorder of the number of questions raised and by whom

### Time Required

As long as it takes for the first person to achieve the game objective (approximately five minutes)

### Game Objective

Be the first participant to correctly articulate the number determined by the volunteer

### Group Size

20 to 30

### Procedure

Select a volunteer. Volunteer chooses a number mentally and says, "I'm thinking of a number." Each participant will get one chance to ask a "yes or no question". No statements are allowed.

Example: Is the number an even number? Is it above 60?

Participants need to keep track of the clues in their heads. The participant who guesses the number correctly with the least number of questions receives a small prize.

Participants who ask repeat questions will be made to do a forfeit, since they have not been listening to the discussion.

### Possible Variations

Using questions like "Is the number greater/smaller than...?" will make it too easy to identify the number. Restrict each type of question to be asked only once throughout the entire game. So, we can ask questions like:

– Is it a 2-digit number?
– Is there a digit "1" in the number?

NOTE: For advanced players only. Success not guaranteed.

### Process.

- What feelings did you experience throughout the game?
- Question for volunteer: What were some of your thoughts as they were asking you the questions?
- What facets of good communication were demonstrated here?

### Practical Application

- What does this activity tell you about communication with another individual?
- Why do we have so many problems with being patient?

## Game 3: "Beengo!"

1. Facilitator calls out number between 1 and 75.

~, 70!

2. Groups circle that number on Beengo card.

| 2 | 62 | 12 | 25 | 61 |
|---|----|----|----|----|
| 10 | 66 | 38 | 75 | 9 |
| 73 | 4 | 22 | 43 | 8 |
| 70 | 7 | 37 | 19 | 61 |
| 72 | 54 | 36 | 16 | 40 |

3. When numbers in a line are formed, a letter is gotten.

 B    BE

4. First group to form the word "Beengo" wins.

BEENGO!

# Leadership Game 3: "Beengo!"

*Key Leadership Understanding*

Leaders seek to fulfill the goals set for the team. Leaders direct the team's strategy.

*Math/Science Concepts Applicable*

Multiplication and division

*Equipment/Logistics*

Beengo card (five rows with five squares per row)
Marker pen
Pieces of papers with simple multiplication and division expressions

*Time Required*

20 minutes

*Game Objective*

Be the first group to have all three team players hit "Beengo!"

*Group Size*

Four to five

*Procedure*

There are four Beengo cards for the four players in a team. On each Beengo card, there are numbers ranging from 1 to 75. The numbers on each card for each of the four players are different.

The facilitator gives a multiplication and division equation. All the four players will try to solve the equation. The player with the answer circles the number on the Beengo card. When a player has answers forming a row either horizontally, vertically, diagonally or with four corners, he or she has got a "Beengo!"

The winning group is the group that manages to have all three of their players hit "Beengo!"

*Process (Questions for group leader)*

- What did you have to do to ensure that everyone in your group completes the Beengo card as fast as possible?

*Practical Application*

- What does this activity tell us about working in teams?
- How are you setting your course for the team?
- What are you doing to build team unity and confidence?

## Game 4: Symmetry

Take as many photos of symmetrical objects as possible.

# Leadership Game 4: Symmetry

*Key Leadership Understanding*

Leaders act with integrity and model the way. They show a good example for others.

Leaders' behaviour reflects the type of leader they are. As seen in symmetrical objects—there is similarity and perfection in the reflection.

*Math/Science Concepts Applicable*

Symmetry

*Equipment/Logistics*

Any location

*Time Required*

20 minutes

*Game Objective*

Be the first group to find (and take pictures of) a fixed number of symmetrical objects within a time limit

*Group Size*

Three to four

*Procedure*

Divide participants into groups of three. Give each group a digital camera, or ensure that each group has a camera phone. Provide a time limit of

20 minutes for the groups to search in the compound for symmetrical objects—architectural designs, rooms, ornaments, etc. They are to take pictures of the symmetrical objects with their camera.

## Possible Variations

Bonus points may be given if players are able to find symmetrical objects of a particular shape—e.g. cylindrical, pyramidal, etc. (up to the facilitator's discretion).

## Process

- Was it difficult/easy to find symmetrical objects in school?
- What were your group's initial feelings when embarking on the activity?

## Practical Application

- Why is symmetry important in our world today?
- What do symmetry and leadership have to do with each other?
- If our behaviour reflects our character, and role modeling is important, what challenges are we facing where our integrity might be questioned?
- What is the behavioural or ethical code that most people of your age live by?

**Game 5: 7 up with a Twist!**

1. Numbers have to be said in ascending order when going clockwise.

2. Numbers to be said in descending order when going anticlockwise.

3. Skip any number that either (a) has a 7 as one of its digits, or (b) is a multiple of 7!

... 13, 15, 16 — (note that they skipped nos 14, 17 and 21.) 18, 22, 20, 19

If you say the wrong number, you get to do a forfeit!

# Leadership Game 5: 7 up with a Twist!

*Key Leadership Understanding*

Leaders listen actively and are disciplined in their approach.

*Math/Science Concepts Applicable*

Multiplication

*Equipment/Logistics*

Two game masters to monitor each group

*Time Required*

15 minutes

*Game Objective*

Be the first group to finish counting the multiples of 7 right up to 98 (either in ascending or descending order).

*Group Size*

Five to Six

*Procedure*

Two groups of players will compete. Each player in the group will take turns to count in multiples of 7 within 0 to 100 (7, 14, …, 98), in ascending or descending order. The respective group's game master will require the group to start all over again for any mistakes or hesitation that the group members make.

*Process*

- What was your main difficulty in counting in multiples?
- Did the process of counting in multiples help you or hinder you? Why?
- How did you feel when you had to recount or start all over?

*Practical Application*

- What lessons do you learn about leadership from this game?
- Were there times when you felt like giving up in your project?

## Game 6: Aladdin's Magic Carpet

Everyone stands on a carpet and tries to flip it over without anyone stepping off the carpet.

1.

2.

"After he crosses, it is your turn..."

3.

4.

# Leadership Game 6: Aladdin's Magic Carpet

*Key Leadership Understanding*

Integrity in leaders is all that matters.

*Math/Science Concepts Applicable*

Geometry

*Equipment/Logistics*

One large piece of groundsheet/newspaper/cardboard to function as the "carpet"

*Time Required*

15 minutes

*Game Objective*

Be the group to flip the "carpet" without anyone stepping out of it in the fastest time possible

*Group Size*

10 to 12 (highly dependent on the size of "carpet").

*Procedure*

All members of the group have to stand on the "magic carpet". The aim of the game is to find a way to flip the "carpet" over without anyone stepping out of it in the quickest time. Once any member steps out of the "magic carpet", the group will have to start all over again.

## Possible Variations

Teams can be given certain restrictions, such as not being allowed to speak or being blindfolded during the task.

Space limits can also be factored into the game. For e.g., game instructors can reduce the amount of space allowed for team members to stand on by tearing the "carpet" after a certain duration of time.

## Process

- Did anyone view the task as impossible?
- How did team members try to keep everyone on the carpet?
- Did the group cheat?

## Practical Application

- What situations exist when our integrity is challenged?
- When was the last time you blundered and had your integrity challenged?
- As leaders, one should have the integrity to keep to the rules of the game, and admit our own mistakes. When was the last time you were challenged to be honest?
- What actions would we take when our integrity is challenged?

## Game 7: Line Up!

1. Everyone stands on a bench, blindfolded. Facilitator shouts a command.

   – Line up according to height!

2.

Everyone reorders himself or herself according to height.

3.

4.

# Leadership Game 7: Line Up!

### Key Leadership Understanding

Leaders give clear and precise instructions.

Team members need to co-operate with their leaders by listening and carrying out their instructions.

### Math/Science Concepts Applicable

Measurement

### Equipment/Logistics

Blindfolds

### Time Required

30 minutes

### Game Objective

Be the first group to line up according to given criteria

### Group Size

10 to 12 per bench

### Procedure

Group members will be blindfolded for this activity.

The group is expected to line up according to height, in ascending order, i.e., from the shortest in front to the tallest at the back. Thereafter they will be asked to do it the other way round, i.e., from the tallest in front to the shortest at the back.

The fastest group to line up accordingly wins.

## Possible Variations

The group can be asked to line up according to other criteria such as birthdates, weight, age, etc.

The group can also be asked to line up according to a given set of criteria without being blindfolded but are forbidden to talk.

Alternatively, the group can be told to line up on an elevated platform (e.g., a bench). They will have to restart the game if any member happens to fall off the platform.

## Process

- How vulnerable did you feel while being blindfolded?
- Was there anyone who was willing to lead the group?

## Practical Application

- What have you learnt about leadership in this game?
- What happens when you have more than one member trying to lead the group?
- How do you ensure accountability by team members?

**Game 8: In Step with the Times!**

1. Each person is given 2 pieces of newspaper.
2. As a group, try to get from point A to point B by stepping on them.

3. The group moves forward by passing newspaper from the back to the front.

4. When the group reaches point B, do the same thing back to point A!

# Leadership Game 8: In Step with the Times!

*Key Leadership Understanding*

Leaders are life long learners. They stay current in the news. They look for trends and are open to new approaches.

*Math/Science Concepts Applicable*

Distance

*Equipment/Logistics*

Newspapers (preferably with pictures)

*Time Required*

10 minutes

*Game Objective*

Be the first group to cross the finishing line

*Group Size*

Six to eight

*Procedure*

The goal of this game is to get from point A to point B, stepping on newspaper.

Divide the group into two teams. Each player is given two pieces of newspaper to move from a starting point A to the finishing line B. To move from A to B, the player has to use the two pieces of newspaper. He or she places one piece on the ground, steps on it, places the other piece before he or she makes the next step. Basically all movements have to be made on the pieces of newspaper. The process continues until the player reaches point B after which the next player from the team will repeat the whole process.

The team with all the players at point B in the shortest period of time wins!

NOTE
When the players are done, the facilitator will collect back all the newspapers and ask the players with regard to the content of the newspapers.

*Possible Variation*

After reaching point B, the whole team repeats the same process to get back to A.

*Process*

- How did you feel about stepping on newspapers?
- Did anything on the pieces of newspaper manage to catch your attention?

*Practical Application*

- How much attention do you pay to current affairs?
- How are newspapers and the media important in our lives?
- How are you developing your skills and confidence?
- How important is it for leaders to be observant and careful in the things they do?

## Game 9: Whose Bubble Is It?

Try to blow the biggest bubble possible.

# Leadership Game 9: Whose Bubble Is It?

*Key Leadership Understanding*
Leaders give hope to others. They are optimistic people.

*Math/Science Concepts Applicable*
Circumference of circle

*Equipment/Logistics*
Chewing gums
Chairs for players

*Time Required*
20 minutes

*Game Objective*
Be the group to produce the biggest bubble or have the greatest number of team players seated

*Group Size*
Six to eight

*Procedure*
Players are divided into groups. Provide each player with a piece of gum and a chair. Each group forms a circle with teammates facing each other.

When the facilitator commences the game, everyone (in standing position) tries to blow a bubble. Each player gets to sit after blowing a bubble. Within a certain time, the group with the greatest number of members seated wins!

Alternatively, the group with the biggest bubble wins.

*Process*

- How did you feel each time someone sat down?
- How did you feel when your bubble burst?
- After you get seated, how did you feel when others in your team are still trying to blow their bubble?

*Practical Application*

- Have you experienced seeing your vision or hopes for the team collapse?
- How can you inject hope into the team when a cause seems lost?

### Game 10: Number Game

1. Each person is allocated a number.

2. Only 1 person in the group is allowed to talk; everyone is blindfolded.

3. Facilitator will shout a command for group members to follow:

   "I want all prime numbers to sit down in a group!"

4. The group has to follow the facilitator's command as fast as possible.

   "I will grab your hand. If you are a prime number, shake it."

5. The person who can talk has the job of coordinating everyone!

# Leadership Game 10: Number Game

*Key Leadership Understanding*

Leaders solve problems and think out of the box.

Leaders listen and are good followers.

*Math/Science Concepts Applicable*

Knowledge of different types of numbers e.g., prime numbers, etc.

*Equipment/Logistics*

Blindfolds

*Time Required*

20 minutes

*Game Objective*

Be the first group to finish the obstacle course

*Group Size*

Six to eight

*Procedure*

All the players are blindfolded. Each player is allocated a number by the facilitator. The numbering relationship (e.g., consecutive numbers, prime numbers, odd numbers, even numbers, negative numbers, in multiples, etc.) is decided by the facilitator.

Only one player in the group called the coordinator is allowed to talk. The coordinator will line team members up in ascending or descending order according to the facilitator's instruction.

The main challenge is for the coordinating player who can talk to communicate with the rest so that he or she can arrange the group members according to the facilitator's instructions.

*Process*

- Was one leader or coordinator sufficient?
- Did communication play a huge role in the group? How?
- Why do you think you could or could not complete this activity?

*Practical Application*

- Why is listening so important?
- How is a leader sometimes a follower?
- How can one be a good follower?

**Game 11: All on A Square!**

Try to squeeze as many people on the mat as possible!

"Squeeze!"

# Leadership Game 11: All on a Square!

*Key Leadership Understanding*

Leaders are confident risk takers. They are able to adapt to changes.

*Math/Science Concepts Applicable*

Estimation of area

*Time Required*

15 to 20 minutes

*Equipment/Logistics*

A large square mat

*Game Objective*

Be the group to keep group members within the given mat

*Group Size*

Five to eight, depending on the size of mat

*Procedure*

Each group is given a mat of the same size.

All the members of the group will start the game by standing on the mat. The facilitator will assign a smaller mat area by, for example, asking the team to fold the mat into half. The group will be asked to set a target above the minimum stipulated by the facilitator as to the number of members which they think can fit onto the mat. Within a given time, the team will then seek to achieve their target.

No parts of their bodies are allowed to go beyond the area of the mat. If they reach their target, 10 points can be awarded for each team. The team with the greatest number of players will be given extra points, but no points will be awarded if the team fails to meet its target (this is to discourage teams from making unrealistic targets without a proper estimation).

After each round, the facilitator can proceed to continue reducing the size of the mat.

### Possible Variations

The mat can be circular, triangular or rectangular to make the game interesting.

### Process

- What was the main difficulty in this game?
- Was everyone comfortable with playing the game? If not, what could be done?
- In what way did this game demand cooperation among members?

### Practical Application

- When was the last occasion when you felt uncomfortable by changes?
- How did you overcome this challenge of change?

Tap the numbers in ascending or descending order.

Only one person is allowed to be in the circle at any one time.

# Leadership Game 12: Station Omega

*Key Leadership Understanding*

Teamwork helps to fulfill the vision of the team.

Leaders allow others to act by giving them opportunities to demonstrate their strengths.

*Math & Science Concepts Applicable*

Knowledge of mathematical concepts, such as perfect squares and prime numbers

*Equipment/Logistics*

Chalk
Stopwatch
A boundary with 50 numbers drawn within it in random order

*Time Required*

20 minutes

*Game Objective*

Be the first team to finish tapping

*Group Size*

Five to six

*Procedure*

Each team will be assigned a boundary of numbers. All the team players are to station themselves outside the boundary. The facilitator will provide a number e.g. 1 and the mathematical relationship e.g. prime numbers. Team players will take turns to enter the circle to tap on all the numbers, in ascending order (e.g., 1, 3, 5, 7, etc.), belonging to the mathematical relationship given by the facilitator. The team that completes the task within the shortest time wins the round. The facilitator will then proceed with another number e.g. 2 and mathematical relationship e.g. even numbers.

Only one member is allowed in the boundary to perform each round.

### Possible Variations

There are many mathematical relationships that the facilitator can provide e.g. perfect squares, multiples, etc. To increase the level of difficulty, team members may be asked to tap in descending or ascending order, in which case the facilitator will need to start with the biggest number.

Alternatively, alphabets can be used instead of numbers in the boundary, where players are asked to spell out scientific terms forward or backwards.

### Process

- How could you have been more effective in this activity?
- What were some of your strengths and/or weaknesses?
- Was teamwork evident in the game?

### Practical Application

- Do we allow others to show their strengths?
- How do we help others overcome their weaknesses?

**Game 13: Shape Me!**

Come up with as many ways as possible to line 10 people to form 4 lines with 4 people each?

e.g.

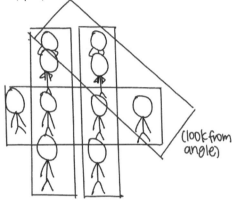

(look from angle)

# Leadership Game 13: Shape Me!

*Key Leadership Understanding*

Leaders overturn assumptions and think out of the box to resolve problems.

*Math/Science Concepts Applicable*

Addition, subtraction, multiplication and division

*Equipment/Logistics*

Paper and markers for each group

*Time Required*

20 minutes

*Game Objective*

Be the first group to come up with as many ways to do the formation(s) within a given duration

*Group Size*

10

*Procedure*

Each group of 10 has to explore as many ways as possible to form four lines such that there are four persons in each line using all the 10 members.

*Process*

- What were some challenges that you face in this activity?

- How did you overcome these challenges?
- What was needed in the group to think of the formation(s)?

*Practical Application*

- What response do you often have when people tell you that you can do better?
- Why is it important to have a different perception of things when one is engaged in a project/group discussion?

**Game 14: Multi-Division**

1. Foot-soldier

'Attention' pose.

2. Calvary

3. Cannon Volley

4. War galley

5. Fortress

a) b)

"Stamp!"

# Leadership Game 14: Multi-Division

*Key Leadership Understanding*

Nurturing and expanding the pool of leaders in an organisation is an important process to ensure continuity.

*Math/Science Concepts Applicable*

Division and number bases

*Equipment/Logistics*

A big open space

*Time Required*

20 minutes

*Game Objective*

Be the one to remain in the game

*Group Size*

20

*Procedure*

Demonstrate the following five commands to the group.

Foot-soldier: Stand at attention
Calvary: Two players stand side-by-side, horse-riding
Cannon volley: Two players kneel on the ground, and a third player leaning on their shoulders

War galley: Four players in a row, "rowing" a boat
Fortress: Five players stand back to back, stamping their feet

Whichever command is given by the facilitator, the players will have to demonstrate these commands. For example, if the command is "War gallery", players will have to arrange themselves in groups of four before they demonstrate the command.

Players who do not follow the commands successfully are pulled out of the game. The last player to remain in the game wins!

### Possible Variations

The actions and commands can be changed and manipulated according to the event or theme.

### Process

- How did you involve others to stay in the game?
- How did you feel about needing others to stay in the game?
- What did you do to increase or reduce your group size according to the commands given?

### Practical Application

- What strategies do you employ to expand your group size?
- How do you persuade or recruit others to join your cause?
- How are we multiplying and growing the strength of our organisation?

### Game 15: Another PIE Jam

Try to get all the 'X's to switch places with all the 'O's.

This can be done through either (a) moving to an adjacent seat or (b) leapfrogging over another person.

# Leadership Game 15: Another PIE* Jam

*Key Leadership Understanding*

Leaders sometimes step back and allow others to proceed first.

Leaders are creative in problem solving.

*Math/Science Concepts Applicable*

Problem solving

*Equipment/Logistics*

Open space

*Time Required*

20 minutes

*Game Objective*

To swap places with the other team

*Group Size*

Two groups of four to five participants each

---

*Pan Island Expressway. The name of Game 15 is contextually Singaporean, with reference to one of Singapore's better known expressways.

*Procedure*

| X | X | X | X | | O | O | O | O |
|---|---|---|---|---|---|---|---|---|

Participants in the two teams, "X" and "O", get to line up as shown. All the participants marked "X" can only move to the right, and those marked "O" can only move to the left. There can be movement only when the adjacent square is empty or that the square beyond the adjacent square is empty, for which the player can "hop" to. Participants "X" and "O" have to figure out a way to mentally swap their places in the end.

*Process*

- How did you feel having to step back when you tried to forge ahead?
- How did you feel about this challenge of stepping into the space of another person?

*Practical Application*

- What does this activity show you about team work and leadership?
- What actions do people make when they get impatient?
- What are we facing right now that is bringing us closer together?
- How can we keep people moving forward in striving for excellence?

# Game 16: Blindfold Polygons

# Leadership Game 16: Blindfold Polygons

*Key Leadership Understanding*

Courage is necessary in any leadership endeavour. Only with courage can leaders overcome difficulties.

Trust is also crucial. Leaders must trust and depend on each other.

*Math/Science Concepts Applicable*

Shapes and sizes

*Equipment/Logistics*

One long-looped string
Blindfolds

*Time Required*

30 minutes

*Game Objective*

Be the fastest group to form the required number of shapes

*Group Size*

Six to eight

*Procedure*

Participants are blindfolded, while holding part of a long-looped string. Without saying any words, they must use the string to form some simple shapes (e.g., hexagon, octagon, triangle, squares, circles, etc.), as required by the facilitator.

Demerit points may be given if the participants request for permission for one of their team members to see or speak.

*Process*

- How did you feel when you were blindfolded?
- How many of you attempted to sneak in a view?
- What were the difficult challenges in this activity?
- How did you overcome the challenges?
- How did you feel when you could not speak to help your team even when you wanted to?

*Practical Application*

- What situations around us are we avoiding or afraid to confront?

## Game 17: Money Mind

50 Math & Science Games for Leadership

# Leadership Game 17: Money Mind

*Key Leadership Understanding*

Leaders are valuable assets. Leaders are principle-centred.

*Math/Science Concepts Applicable*

Addition

*Equipment/Logistics*

10 auction items representing each of the following leadership qualities—trust, courage, responsibility, respect, diligence, courage, perseverance, humility, confidence and compassion

*Time Required*

20 minutes

*Game Objective*

Bid for as many auction items with the given amount of money

*Group Size*

Five to six

*Procedure*

Each auction item represents a leadership trait (e.g., trust, courage, etc.). Give each group a pre-determined amount of money. Each group will have to decide on which auction item (or leadership trait) they want to bid and for how much.

This exercise will get the group to decide which value they believe to be most important to them. Their debate about values will also mean that the group might be divided over how their monetary resources are used.

*Process*

- How did you feel when you lost a value?
- How did the group decide which value was important?

*Practical Application*

- What does this activity tell you about values in a leader?
- Are values important to a leader? Why or why not?
- Which is the value that we hold onto in trying situations?
- What is the code that you live by?

## Game 18: Momentum

This is a multiplication game that speeds up as it goes on.

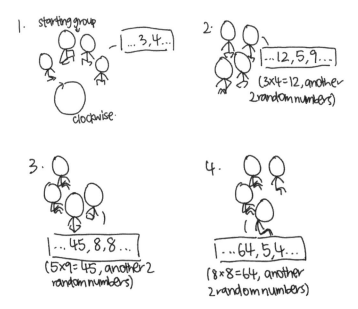

1. starting group
   -|...3,4...|
   clockwise.

2. |...12,5,9...|
   (3x4=12, another 2 random numbers)

3. |...45,8,8...|
   (5x9=45, another 2 random numbers)

4. |...64,5,4...|
   (8x8=64, another 2 random numbers)

# Leadership Game 18: Momentum

*Key Leadership Understanding*

Leaders are not afraid to take charge. They energize the team and keep up the momentum.

*Equipment/Logistics*

Nil

*Math/Science Concepts Applicable*

Simple multiplication

*Time Required*

30 minutes

*Game Objective*

Be the last group to remain in the circle

*Group Size*

Five to eight

*Prior Preparation*

Nil

*Procedure*

Have each group sit together in a "circle", such that each group can see other groups.

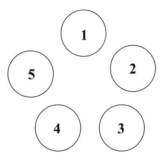

The facilitator will start by providing two numbers between 1 to 10 (e.g., 5 and 9). The first group will then have to say "45" (because 5 x 9 is 45) and pick another two random numbers between 1 to 10 (e.g., 2 and 4). The 2nd group will have to answer "8", and so on. They are to answer immediately after the 1st group finishes, so it becomes like a rhythm. In this example, what you will hear is something like: 5, 9, 45, 2, 4, 8, etc.

The 1st group which breaks that rhythm is disqualified from the game, or they will have to do a forfeit to stay in the game.

The catch is that whatever it is, the group must chant the numbers together in unison. If one member chants the wrong answer, the group's rhythm will be broken as well.

*Possible Variations*

The game can be sped up to increase the difficulty level.

*Process*

- How did you feel when one group member's error caused the whole team to falter?
- How would you do this task differently the next time?

*Practical Application*

- What are the biggest difficulties in taking action when you face high levels of emotions like fear?
- What kind of situations call for encouragement to be given and how can we best do it?
- How can we press things forward?
- What lessons can you gather about leaders and teamwork?

## Game 19: Graphs

1. Come up with a survey question and poll the class about it.

"How many people have an allowance of more than $20 a week?"

2. Graph your results and present them to the rest of the class.

"There is a positive correlation between your allowance and your grades!"

# Leadership Game 19: Graphs

*Key Leadership Understanding*

Leaders know their fellow leaders and teammates well. They know each other's strengths to synergize them.

*Math/Science Concepts Applicable*

Graphing skills, tabulation and data interpretation

*Equipment/Logistics*

Graph paper and mahjong paper
Stationery
Computer
Projector
Microsoft Excel

*Time Required*

About one hour

*Game Objective*

To find out about the preferences of others and accept the differences without unfair judgment

*Group Size*

Three to four

*Procedure*

Each group member comes up with a question that he or she wants to ask the teammates. Questions asked should elicit responses with regard to team members' interests and preferences—questions concerning eye colour, birth dates, and number of family members, etc. Members are encouraged to be as creative as possible in the types of questions they ask and how they tabulate their responses.

Groups are to present their results in the form of graphs to be presented to the other groups. Groups are challenged to show their results in a creative way (e.g., if everyone's favourite song is the national anthem*, they could sing it during the presentation, or even develop a skit/play/musical to present the data).

### Process

- What was fun about this activity?
- What did you learn about your fellow teammates?
- What facets of good communication did you see being demonstrated here?

### Practical Application

- What could be done to improve the communications and get things done more quickly and efficiently?

---

*Players are to be sensitive about the national anthems of their group members.

## Game 20: Angle of Elevation

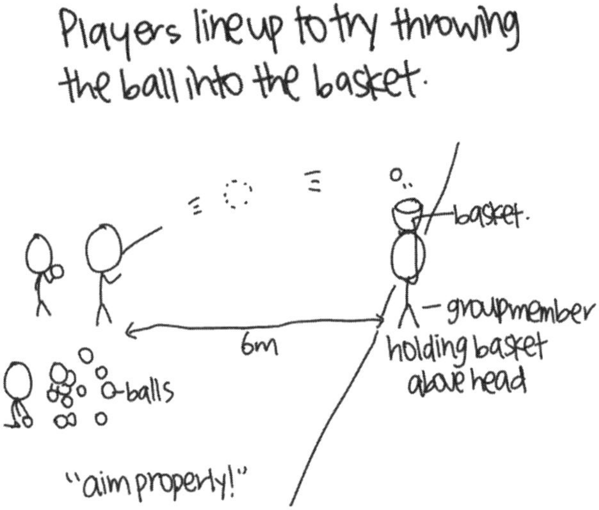

Players line up to try throwing
the ball into the basket.

basket.

— groupmember
holding basket
above head

6m

o-balls

"aim properly!"

# Leadership Game 20: Angle of Elevation

*Key Leadership Understanding*

Leaders are responsible for all members in the team.

Leaders are open to diverse viewpoints.

*Math/Science Concepts Applicable*

Distance and angles of elevation

*Equipment/Logistics*

100 coloured balls and one basket per team

*Time Required*

20 minutes

*Game Objective*

Be the group with the most number of balls in the basket

*Group Size*

Five to six

*Procedure*

Players are divided into two groups. The goal of each group is to get coloured balls into their basket.

Have two players hold their basket on their heads. They are allowed to tilt or lift the baskets to try to catch the balls which their team members toss.

The thrower and the basket handlers are about six metres away from each other. (Draw a line to demarcate the distance.)

All the coloured balls have to be tossed in the given time. The team with the most number of balls in the basket is the winner.

*Possible Variations*

Only overhand throws or underhand throws are allowed.

*Process*

- How did you feel throughout the game?
- How did you (basket holders) feel when some of the balls missed your basket or hit you?

*Practical Application*

- If each coloured ball refers to a team member with a different viewpoint, how would you respond to this team member?
- If each coloured ball refers to a team member, how would you feel about the loss of each team member?
- How can we demonstrate a general respect for each other's diverse talents?

**Game 21: Danger Zone!**

Guide the blindfolded people across the obstacle course

LEFT 2 STEPS!

Navigating the maze

Everytime they touch an obstacle, they will be given three simultaneous equations.

Solving simultaneous equations at the end

50 Math & Science Games for Leadership

# Leadership Game 21: Danger Zone!

*Key Leadership Understanding*

Leaders have visions and communicate them to others.

*Math/Science Concepts Applicable*

Simultaneous equations

*Time Required*

30 minutes

*Equipment/Logistics*

Ropes or raffia string to mark out out-of-bound areas which serve as obstacles in the "mine field"
Blindfolds
Chairs, tables, or any furniture as large obstacles
Outdoor or open space

*Game Objective*

Be the first group to complete the task which includes crossing over a "mine field" and solving simultaneous equation

*Group Size*

Eight to 15

*Procedure*

The game can be played indoors (in a room), or outdoors (a field or open spaces).

Before the activity is conducted, the facilitator sets up the entire game area, positioning some chairs and tables as obstacles and marking out the out-of-bound areas to create a "mine field". Facilitators are to be stationed inside the game area throughout the game to ensure the team does not cheat, and reinforce safety.

Each group is given about five minutes to discuss and develop a game strategy. Everyone will then be blindfolded except for one member which the team selects to be allowed to see in the game. No one including the "seeing" player is allowed to speak. The "seeing" member will guide each blindfolded member across the "mine field".

There is a time limit set for the entire team to cross the "mine field". The "seeing" member has to ensure that no one touches the obstacles or steps on out-of-bound areas set aside as "mines". Every time a player touches an obstacle or steps on an out-of-bound area, the player will be given three simultaneous equations to solve as a penalty after the entire team crosses the "mine field".

Once the entire team makes it through the "mine field", they are then allowed to take off their blindfolds and work on the simultaneous equations that all the team members have accumulated. The game ends when all the equations are solved.

*Possible Variations*

Background music can be played loudly (if indoor) to distract the team in the game.
Some or all the group members are permitted to speak.

*Process*

- How difficult was it to cross the "mine field" without vision?
- How critical was it to have a guiding voice?
- How did you (as the "seeing" member) successfully guide your team members across the "mine field"?

*Practical Application*

- When was the last occasion you felt inspired by a leader's vision?
- What was your last "mine field"? How did you overcome it?

**Game 22: Fishing**

"aim for the parentheses!"

ADVANCED

# Leadership Game 22: Fishing

*Key Leadership Understanding*

Leaders have to balance between being pragmatists and idealists.

*Math/Science Concepts Applicable*

Basic addition, subtraction, multiplication, division and physics for the building of fishing device.

*Equipment/Logistics*

Drinking straws
Scotch tape
Metal bendable wires
Tags with numbers and mathematical symbols and signs
Rings to be attached to tags for hooking purposes
Judge to decide on complexity of equation

*Time Required*

One hour or more

*Game Objective*

Be the first group to create the most complicated equation

*Group Size*

Four to five

*Procedure*

A boundary is created to place all the tags. The tags will have numbers or mathematical symbols and signs (addition, subtraction, division, multiplication, differentiation, bracket, equal or percentage sign, etc.). Ensure that a combination of the tags can be used to form mathematical equations (e.g., the five tags – 13, 17, –, +, and 30 – can be used to form equations such as "30 – 17 = 13").

Each group is given metal bendable wires, rings and scotch tape. They have to use these items to build a hooking device for fishing out the tags to build their mathematical equations. They can only fish from a particular distance.

Within the given time and regardless of the number of equations they come up with, the winning group is the one with the equation considered most complex by the judge.

### Possible Variations
Teams are allowed to trade tags if they wish to.

### Process
- Did you succumb to the pressure of winning and completing the game in the time given, compromising on the complexity required?
- Did you encourage those fishing out the tags which is not an easy task?

### Practical Application
- What is your stand on issues in your present organisation if the costs were high?
- How can you achieve high standards without overlooking the practical aspects?

## Game 23: The Olympics of Science and Math

1. Fork throw / cotton ball putt

"throw further!"

2. Water sponge squeeze

"squeeze harder!"

3. Balloon relay

How far do you think you can go?

How accurate were your estimates?

# Leadership Game 23:
# The Olympics of Science and Math

*Key Leadership Understanding*

Leaders go the distance. They take on an Olympian spirit and perform at their best.

*Math/Science Concepts Applicable*

Weight, volume and distance

*Equipment/Logistics*

Work sheet listing the "Olympic" events
Scales
Marbles
Cotton balls
Straws
Paper plates
Rulers
Sponges
Water containers
Small boxes
Coins

*Time Required*

About one hour

*Game Objective*

Be the group that accumulates the best total score from all the events

One to three

*Procedure*

This activity is a series of mini-games revolving around the concept of measurement, estimation and approximation.

Possible mini-games include fork throw, water sponge squeeze, cotton ball putt and relay. The players are given a sheet listing the "Olympic" events, of which they will set a goal by estimation.

Examples include distance fork can reach, amount of time needed to do the relay, etc. They will then rotate among the various stations to measure how close they are to their estimate. The player with the most accurate estimates will be the gold medalist. Standard metric or imperial units may be used.

*Process*

- How did you feel participating in these events?
- Did you try your best in your event?
- How do you think you have measured up to the goals that you have set?

*Practical Application*

- How can you exercise leadership with limited resources, support and power?
- What level of commitment and action do people see from you in your organisation?

## Game 24: Vector Pull

Try to move the bottle into the mark in the middle of the circle by shifting the raffia tied to it.

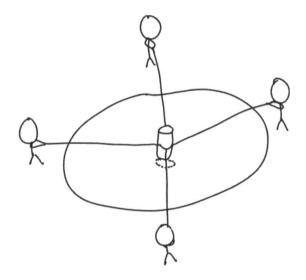

"Move a bit more to your left!"

# Leadership Game 24: Vector Pull

*Key Leadership Understanding*

A leader listens to every member and searches for the best way forward.

*Math/Science Concepts Applicable*

Vectors

*Time Required*

20 minutes

*Equipment/Logistics*

Four ropes or pieces of string
Raffia string to create boundary
Bottle/ring
Chalk

*Game Objective*

Be the group to move the bottle into the designated area within the shortest time possible

*Group Size*

Four to 10

*Procedure*

Tie the four strings to the bottle and use some raffia to create a boundary around the bottle.

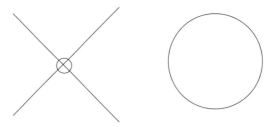

Draw out a marking within the circle using chalk, where the bottle is to be shifted to.

Using the four strings and not stepping into the area bounded by the circle, try to move the bottle into the centre of the circle as marked out in the shortest time possible. The bottle must not topple over.

*Possible Variations*

Other items, aside from a bottle, can be used

**Process**

- How did you cooperate as a team to move the bottle in the direction you wanted?
- Was there anyone who stepped up as a leader to direct the group?
- Were there times when you felt that the task was impossible? Why?

*Practical Application*

- How can we keep people moving forward to achieve a goal?
- Have you experienced a situation in a group task when everyone talked and nobody cared to listen?

# PART II

# SCIENCE GAMES

### Game 25: Rubber Banding

rubber band

"6 more to go!"

Transfer the rubber bands from one end to another in the shortest time possible by putting a straw in their mouth and using it to transport.

# Leadership Game 25: Rubber Banding

*Key Leadership Understanding*

Leaders inspire others around them to work together to achieve a common goal.

*Math/Science Concepts Applicable*

Centre of gravity

*Equipment/Logistics*

Rubber bands
Straws

*Time Required*

20 minutes

*Game Objective*

Be the group to transfer a fixed number of rubber bands while standing in a line within the shortest time possible.

*Group Size*

About six to eight (but can be changed accordingly)

*Procedure*

Each group member is supposed to have a straw in his or her mouth and transfer all the rubber bands down the line as a group. The group is supposed to accomplish the task within the shortest time possible without using their hands to support the straws or come in contact with the rubber bands.

## Possible Variations

Instead of transferring rubber bands, the group could instead transfer table tennis balls using plastic spoons. The concept is similar but it is definitely harder!

## Process

- What difficulties did you encounter throughout this activity?
- How did you feel when the rubber band dropped?
- Did you feel comfortable about this activity? Why?
- How could you improve the next time?
- Could you work within the limited preparation time?

## Practical Application

- What can we do to move our team forward?

## Game 26: Biodiversity

table salt!

Each balloon will have the name of a chemical compound on it.

When that chemical compound is called, make sure that the balloon with that chemical compound is thrown higher up than the rest.

# Leadership Game 26: Biodiversity

*Key Leadership Understanding*

Leaders identify different groups and individuals and support diverse community.

*Math/Science Concepts Applicable*

Biodiversity and ecosystem

*Equipment/Logistics*

100 multi-coloured blown up balloons
Marker pens
20 balloons per group of 10 participants

*Time required*

20 minutes

*Game Objective*

Be the last group to keep afloat balloons with the compound names on them

*Group Size*

10

*Procedure*

Assemble the group in one large circle. Have balloons with the chemical compounds written on them. Balloons of different colours are thrown into the circle. The group has to continue tossing the balloons to keep the balloons afloat. Whenever the compound name is mentioned, the group will have to make sure that the balloon is tossed higher than the rest. If the compound balloon touches the floor, the group forfeits the game.

*Process*

- How did the activity make you feel? Was it challenging to keep the coloured compound balloons afloat? Why?

*Practical Application*

- Imagine if the coloured compound balloons were foreign teammates in your group. Did the support these coloured compound balloons got in the activity reflect the support the foreign teammates get in your organisation? Did any of the balloons burst? How is that similar to peer-pressure given in opposing directions?
- What should group leaders do to support foreign group members?
- What can we do to make sure that our campus or leadership group does not become elitist or divisive?
- What are the reasons we use to reject people or refuse them an opportunity to participate?

**Game 27: Boeing or Airbus?**

1. Each team is given 2 sheets of A4 paper to fold paper planes with.

2. The objective is to fold the paper plane such that it flies the longest distance possible

3. After folding, each team sends a representative to the starting point to fly or release the aeroplane.

21, 22, 23 ...

# Leadership Game 27: Boeing or Airbus?

*Key Leadership Understanding*

Leaders set the course for others. Leaders soar with a well-designed plan.

*Math/Science Concepts Applicable*

Aerodynamics; Knowledge of streamline body and airflow; Knowledge of the projectile motion

*Equipment/Logistics*

A4-sized paper
A big open space

*Time Required*

20 minutes

*Game Objective*

Fold a paper plane and fly it in such a way that it covers the longest distance possible.

*Group Size*

Six to eight

*Procedure*

Give each team two pieces of A4-sized paper and let them know the objective of the game. They are then given five minutes to discuss what would be the best way to fold the airplane such that it achieves the objective.

When the time limit is up, one representative from each team will proceed to the same starting point which could be on level ground or on a slope. All the representatives will fly their plane at the same time. The winner is the group whose paper airplane covers the longest distance.

The whole process is repeated so players can modify and improve their folded planes. The overall winner will be the group with the greatest number of wins.

*Possible Variations*

Instead of the longest distance, other indicators (e.g., amount of time it stays airborne, etc.) can be used as well to award points.

Alternatively, we can also have a dartboard where players use the planes to hit various places at the dartboard to try and get points. The closer the plane is to the bull's eye, the higher the points they receive, and the group with the highest number of points wins.

*Process*

- How did you feel about the activity?
- How did the different factors affect the flight of the plane?
- How did you decide which was the best way to fold the airplane such that it would achieve the objective in the best way possible?
- Was there a clear leader who directed the process?

*Practical Application*

- What do we do in the event that our plans go awry?

## Game 28: Mahjong Dominoes!

1. Everyone lines up and takes turns lining dominoes.

2. The dominoes will then be touched and the team with the most number of dominoes that fall wins.

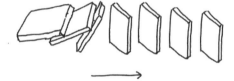

# Leadership Game 28: Mahjong Dominoes!

*Key Leadership Understanding*

Leaders work for the long term benefit of the organisation.

Leaders influence others and make a positive difference.

*Math/Science Concepts Applicable*

Conversion of energy

*Time Required*

40 minutes

*Equipment/Logistics*

A set of mahjong tiles

*Game Objective*

Be the greatest number of tiles

*Group Size*

10 to 15

*Procedure*

Divide the group into two teams. Every player in each team is given four dominoes.

The first player from each team runs to the counter with their four dominoes and sets them upright on the floor (or with the tiles on top of each other) in a straight line. They run back and tag the next team member, who runs forward to do the same. This continues until all members of each team have placed their dominoes upright. The last player in the team will touch to trigger the fall of the dominoes that have been lined up in a straight line.

The team with the greatest number of dominoes that fall wins.

## Possible Variations

Cards can also be used in place of tiles, such that each team forms a card tower.

## Process

- What was the most challenging part in the game?
- How did you feel when the tiles dropped? Did you feel as if your work was demolished?

## Practical Application

- Imagine the wall built is the work of leaders. What lessons do you learn about leaders in this game?
- What kind of high standards are we asking others to maintain?

### Game 29: Broken Telephone Line/Fax Machine

1. Group members stand in a line, spread out from each other.

2. The 1st person will be given a list of scientific terms.

Psst...

3. Try to communicate these terms down to the last person.

# Leadership Game 29:
# Broken Telephone Line/Fax Machine

*Key Leadership Understanding*

Leaders provide the vision and rally support from others.

*Math/Science Concepts Applicable*

Scientific concepts depending on the terms used

*Time Required*

10 to 15 minutes

*Equipment/Logistics*

Cards, each with a different scientific term (e.g., oxidation, apoptosis, etc.) written on it

*Game Objective*

Be the group to provide the most accurate depiction of the term

*Group Size*

Eight to ten

*Procedure*

All players have to line themselves up in one straight line. No talking is allowed.

The players will elect a leader among themselves to go forward and receive a stack of cards with the scientific terms to be communicated. The leader will then have to commit all the terms to memory, and pass on the terms to the next member of his or her group. This process continues with each member receiving and passing on the list of terms to each other.

The last person in the line will have to reproduce the list to the facilitator. This list will be compared against the original list given to the team leader. Points can be awarded based on how similar the end message is to the original.

### Possible Variations

Instead of passing down scientific terms, the leader may receive a picture depicting a scientific process going on. He or she will then have to draw on his team member's back, and this drawing is passed on. The last member will have to guess what scientific process is occurring.

Facilitators must ensure no talking while members are passing the message.

### Process

- What difficulties did you encounter during this activity?
- What did you do when you were unsure of what the message that was being passed down was?
- How did you think you fared when doing this activity? Could you have done better?

### Practical Application

- What facets of good communication do you see being demonstrated here?

### Game 30: Circle of Influence

1. Team members stand in a circle with their feet touching each other's.

2. The opponent tries to roll the ball through the legs of the persons making up the circle.

3. Hands can be used to stop the ball.

4. When the ball goes through, the player is out.

out.

# Leadership Game 30: Circle of Influence

*Key Leadership Understanding*

Leaders build their circle of influence as they become the centre of an individual's life.

*Math/Science Concepts Applicable*

Human reaction time when the environment changes

*Equipment/Logistics*

Soccer balls

*Time Required*

20 minutes

*Game Objective*

Be the group with the least number of balls out of the circle

*Group Size*

Eight to ten

*Procedure*

Have each group of players form a circle with their legs spread out so that their feet are touching those of the players on either side. One player from the opposing team will be in the centre with the ball. The goal of this opposing player is to get the ball out of the circle. Players forming the circle are not allowed to move their legs, but they could stop the ball with their hands from going through their legs. Once the ball gets through a player's legs, that player is out of the game.

As the number of players forming the circle reduces, the circle will get smaller and smaller until the group size is down to two players at which point the game ends.

The game tests the players' reaction time to the ball coming at them from the opposing team player.

*Process*

- How did you feel when the ball went out of the circle?
- How did you feel when you were in the centre of the circle?
- Were you able to help your teammates prevent the ball from getting out of the circle?

*Practical Application*

- What is the value of getting everyone involved and opening up opportunities for greater contribution?
- If the circle represented the leadership circle, how can you strengthen the team each time it is reduced in size?

## Game 31: The World's Your Oyster

1. Each team is given a list of scientific terms.

2. Using whatever resources available, find out the definitions of these terms in the shortest possible time.

# Leadership Game 31: The World's Your Oyster

*Key Leadership Understanding*

Leaders search out of the box to find new things. Challenging the boundaries to improve processes and exploration are key leadership skills.

*Math/Science Concepts Applicable*

Biology terms

*Equipment/Logistics*

Library
The Internet
Various posters and signs
Participants' work
Paper and pencils

*Time Required*

30 minutes

*Game Objective*

Be the first group to figure out the definition of various scientific terms

*Group Size*

Two to three

*Procedure*

Form groups of two to three participants and have the groups search library books, the Internet and every possible source to find out about the scientific terms. Contest could be held to see who is the most successful group in searching for all the categories of words. You could offer some prizes to encourage careful searching.

*Process*

- What feelings did you experience when you did the search?
- What source of information proved most useful for you? Why?

*Practical Application*

- What does this activity show you about the importance of searching as a leadership practice?
- Why do some of us not challenge the boundaries? What situations around us are we avoiding or afraid to confront?
- What are the biggest obstacles to overcome in taking action?

## SOME POSSIBLE/SUGGESTED SCIENTIFIC TERMS

Natural Selection
Femur
Exoskeleton
Endocrine system
Fimbria
Oxytocin
Ecosystem
Ornithology
Zoonosis
Golgi apparatus
Start codon
Phospholipid bilayer
Mimicry
Phagocytosis
Endoparasite
Distal Tubule
Synapse
Carpel
Bacteriophage
Monocotyledon
Beta-Carotene
Apoptosis

# Game 32: Once upon a Time

1. Pick a scientific term from the box.

a)            b)

2. Write a song/story/rap using that scientific term.

3. Perform it!

# Leadership Game 32: Once upon a Time

*Key Leadership Understanding*

A leader sees opportunities and angles where others cannot or do not.

Leaders understand the importance of creativity as a key skill in inspiring the vision for others.

*Math/Science Concepts Applicable*

Scientific terms

*Equipment/Logistics*

Paper and pencils
Collage paper

*Time Required*

30 minutes (15 minutes of preparation and 15 minutes for presentation)

*Game Objective*

Be the group to devise the most creative story with a given scientific term

*Group Size*

Three

*Procedure*

Each group will pick a card with a key scientific term. (See activity in Game 7: The World's Your Oyster.) Each group has to tell a story or write a story/

poem/rap with the use of the term. The winning team will be one rated the most creative by the teacher.

*Process*

- What challenges did you face when devising the story?
- Did you try to do something different? How?
- What does this activity show you about creativity as a leadership trait?

*Practical Application*

- How do we help others to become more creative?
- How are you setting the course for your team as a leader?

**Game 33: Nervous Pulses**

1. Facilitator shows a representative from both groups a number.

2. Representative finds a way to transmit the message to the person at the end of the line without talking.

*clap* x 2

3. Person at the end of the line will take out that specified number of ping pong balls from the bucket.

# Leadership Game 33: Nervous Pulses

### Key Leadership Understanding

Leaders are patient with others. Leaders listen to good advice and communicate accurately.

### Math/Science Concepts Applicable

This is a simulation of how nerves work. Once an organ senses something, the nerves will send a pulse to the brain, and back to the organ or other parts of the body. In this game, the person at the back has received information, and will have to send pulses in any form (except verbally) to pass the information down to the person in front.

### Equipment/Logistics

Ping pong balls

### Time Required

Five minutes per round

### Game Objective

Be the fastest group to pass the message most accurately

### Group Size

10 to 20

### Procedure

Two groups of participants will sit in a line parallel to each other. In between the two rows, a bucket of ping pong balls will be placed. The facilitator will flash a number to the last person of both rows. They will then have to utilize any way except verbal communication to pass down the message to the person in front. When the message is passed down, the person in front has to take the specified number of ping pong balls from the bucket. The fastest and most accurate group (with ping pong balls closest to the number flashed) wins.

To determine the fastest group, facilitators can consider the following:

Assuming the bucket has $n$ number of balls,
the range of numbers should be from $\{[0.5 \times n] + 1\}$ to $n$.

### Process

- What were some factors that determined the group's success or failure?

### Practical Application

- Do you think verbal communication will be more effective in this case? Why?
- What aspects of good communication do you see being demonstrated here?

## Game 34: Group Obstacle Race

1. Finish the obstacle race as fast as possible.

2. Cones or ropes can be moved.

3. Overcome an obstacle...

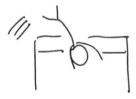

4. ...or bypass, and answer a question in 10 s.

# Leadership Game 34: Group Obstacle Race

*Key Leadership Understanding*

Leaders show perseverance and utilize the strengths of their team members.

*Math/Science Concepts Applicable*

Dependent on science questions posed

*Equipment/Logistics*

Ropes and cones (to be set up)
Strings
Set of science questions

*Time Required*

15 minutes

*Game Objective*

Complete the obstacle race in the fastest time possible

*Group Size*

Eight to nine

*Procedure*

Group members must complete the obstacle race together and cross the finish line together. Ropes can be raised for the whole group to jump across them at one go. Cones can be placed such that the group has to walk in a different manner. Ropes can also be raised for the whole group to crawl underneath them. The whole group must be tied up together with strings so that they can "stick together".

At each obstacle, the group members will have the option of whether they want to bypass it or overcome it. If they choose the former option, they will be given a science question which they must answer within 10 seconds.

Forfeits can be given if group members split up, do not carry out the tasks correctly, or fail to answer the questions.

*Process*

- Did you feel uncomfortable being tied together? Why or why not?
- Was there a leader that could give directions? Or were there too many?
- What was the main difficulty in this game?

*Practical Application*

- How does teamwork come into play in this game?
- Which is the most difficult situation you are currently facing, where it might be easy to get discouraged?
- What takes teams so long to recognise the true extent of their problems?

**Game 35: Scientific Scrabble!**

1. Draw 7 letters.

2. Form words. Only math or science terms allowed.

3. Different letter = different points.

4. Judge will determine if term is valid.

# Leadership Game 35: Scientific Scrabble!

*Key Leadership Understanding*

Leaders learn to strategize with others (teammates within and external parties) to deliver results.

*Math/Science Concepts Applicable*

Mathematical and scientific terms; Addition skills

*Equipment/Logistics*

Scrabble board
Scorer and score sheet
Judge

*Time Required*

One hour

*Game Objective*

Be the group to attain the highest score possible at the end of the game

*Group Size*

Three to four with a total of four groups.

*Procedure*

Each group plays "scientific" scrabble. The game is played as a team rather than as individuals. Like Scrabble, players from each team will draw seven letters from a bag and form words on a scrabble board. The time limit for each turn is five minutes. Each letter has a different numerical score. Therefore, each word will have different summative scores.

Only mathematical or scientific terms are allowed (e.g., *biodiversity, pi, beta-carotene*, etc.). The judge decides whether or not the word used is scientific. Teams are allowed to explain and appeal. However, the judge's decision is final.

The team with the highest score wins.

*Process*

- Were you exasperated when your team members did not take up your suggested formation of words?
- How did you feel when your team won/lost?
- Would you have strategized each formation differently?
- Was communication effective among team members in the game?

*Practical Application*

- What principles are involved in good communication and teamwork?

## Game 36: Murder Mystery

# Leadership Game 36: Murder Mystery

*Key Leadership Understanding*

Teamwork materialises the group's dream or vision.

Leaders need to have a good listening ear and be receptive to the opinions of team members before making any decisions.

*Math/Science Concepts Applicable*

Forensic science and logical deductions

*Equipment/Logistics*

Magnifying glass
Paper
Stamp-pad
Shoe print and thumb print of "murderer"
Shoe print and thumb print of the "victim"
Six or more players as "suspects" (one to function as the "murderer", one as the "victim" and the remaining four as extras).
Six or more hideouts for each of the "suspects"

*Time Required*

One hour or more

*Game Objective*

Be the first group to deduce correctly who is responsible for the "crime"

*Group Size*

Five to six

## Procedure

The main objective of this game is for a group to work together as detectives to solve a murder mystery, using the clues given to them.

The six or more "suspects" will be assigned to different hideout locations. Two sets of thumb and shoe prints (one set belonging to the "murderer" and one set to the "victim") will be given to each group which has to conduct an investigation to trace the "murderer" and "victim". Players have to search out all the "suspects" and devise creative ways to obtain the "suspects'" thumb and shoe prints. Each "suspect" will decide whether the approach is creative enough for him or her to allow the prints to be given to the group.

The fastest group to solve the mystery wins.

## Process

- Was it tiring to carry out the same checks over and over again for the various "suspects"?
- Could this problem be solved or made less problematic?
- Did everyone put in an effort to solve the mystery or was it just a small group doing all the work?
- Did your group work in pairs or as a group when you conducted your investigation with each "suspect"?

## Practical Application

- How much do you contribute to your team?
- Do others consider you a team player?

## Game 37: Food Web

1. Each person is given a card with an organism written on it.

2. With all these organisms, form a food web.

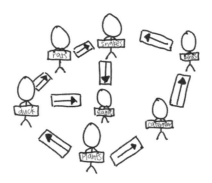

3. If organisms are taken out, there may be a need to redo the food web to fit the circumstances.

# Leadership Game 37: Food Web

*Key Leadership Understanding*

Leaders encourage participation among members within the group and inspire all to work towards a common goal.

Leaders can be both models and followers.

*Math/Science Concepts Applicable*

Ecology

*Equipment/Logistics*

Cards, each with the name of an organism (all the organisms must be related in a food web).
Stopwatch

*Time Required*

15-20 minutes

*Game Objective*

Aim to involve every team member in the food web

*Group Size*

Five to six

*Procedure*

Organism name cards will be distributed to the group members indicating the role each has to play. All the members must be involved to form a gigantic food web. The facilitator has to time how long they take.

Upon completion of the first food web, the facilitator has to proof-read the food web to ensure it is logical. The facilitator will let the group know how long they had taken to complete the first food web.

Next, the facilitator will then remove a couple of "organisms" or members from the food web and get the group to create a food web again. The team will explain to the facilitator the relationships in the food web they have formed.

The aim of the game is to challenge players on how to reposition themselves each time some "organisms" are taken out.

*Process*

- What was the main difficulty in this game?
- Was every member important?
- When a couple of "organisms" were removed from the "food web", what impact did that have on the remaining "organisms"?

*Practical Application*

- In what ways are food webs similar to how a team functions?
- Are we the "predators" or the "prey" in our organisation?
- When some members are taken away, how does that impact a team? How do we respond under such a circumstance?

### Game 38: Cam the Chemicals!

Take photos of things that best represent the chemical names given to you.

# Leadership Game 38: Cam the Chemicals!

*Key Leadership Understanding*

Leaders bring others into the picture and galvanise the team to achieve results.

*Math/Science Concepts Applicable*

Atomic structures

*Equipment/Logistics*

Piece of paper with molecular or atomic structures written on it
Camera
Laptop

*Time Required*

About one hour

*Game Objective*

Determine the materials that make up a given chemical structure

*Group Size*

Five to six

*Procedure*

Players are to bring their own cameras and laptops. Each group will receive the piece of paper with molecular structures. Players can use the Internet to figure out the materials that make up the molecular structures.

Once they find out what materials the structures represent, they are to go around the compound or environment to scout for these materials to take pictures of them. The first group to finish the task with the least number of mistakes wins.

### Possible Variations

Certain restrictions might be imposed to make the game more challenging. For example, instead of researching on the Internet, they are given only a chemistry textbook.

### Process

- How did you allocate your time given that you had one hour?
- What was the role each member played in the whole activity?
- How did you find your performance both as a group and as an individual player?

### Practical Application

- When you are unsure about the work you have been assigned to, what do you do?
- What situations are you facing right now that require more support?

**Game 39: Taboo!**

1. A player from the group will receive a scientific term with three words that he/she is not allowed to say.

2. He/She has to describe the term to the teammates and they have to try guessing the term.

It has a head and a tail...

??? What can it be?

# Leadership Game 39: Taboo!

*Key Leadership Understanding*

Communicating your vision clearly is a critical aspect of effective leadership.

*Math/Science Concepts Applicable*

Scientific terms and concepts

*Equipment/Logistics*

Cards, each with a scientific term or concept and indicating three (taboo) words which are closely related to the term/concept

*Time Required*

About one minute per player

*Game Objective*

Explain a scientific term or concept without using certain words for teammates to guess the term or concept given

*Group Size*

10 to 20

*Procedure*

Players take turns to receive a card with a scientific term or concept on it. He or she is to describe the term or concept as vividly as possible for the rest of the players to guess at the term. However, the three words written on the card are meant to be taboo and are not to be uttered while describing the term or concept. The player forfeits his or her turn when the taboo word is spoken.

If the player finds the term on the card too difficult to describe, he or she can request for a change up to two times.

*Process*

- All of you had a chance to describe a term. How did you feel when your teammates could not guess correctly?
- When all of you were guessing the term, how did you feel when time was running out and you could not get the correct answer? Did you feel annoyed when repeated answers were given?
- Did you blame your peers for not being descriptive enough or did you encourage them?

*Practical Application*

- What are we doing to build team unity and confidence?

### Game 40: Bob the Builder

1. Pass a test tube containing water from the front of the line down:

2. The last person pours the water into a bucket at the end of the line.

3. He/she runs to the start of the line, and the cycle starts again.

4. Overall setup:

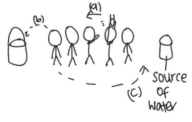

# Leadership Game 40: Bob the Builder

*Key Leadership Understanding*

Leaders build legacies. Leaders never pass an empty mug to the next.

*Math/Science Concepts Applicable*

Physics concepts such as gravity and fluid dynamics

*Equipment/Logistics*

Test tubes
Raffia string
Tape
Rubber bands
Buckets
Water

*Time Required*

30 minutes

*Game Objective*

Be the group that obtains the largest volume of water in the bucket

*Group Size*

Six to eight

*Procedure*

Each group of players is to stand facing each other's back in one line. Each group is given a test tube and a large bucket of water placed at point A.

The goal is to transport as much water as possible from the bucket at point A to point B. Players take turns to scoop the water with the test tube and pass it overhead to the player behind until it gets to the end of the line. The last player to receive the test tube pours the water from the tube into an empty bucket. He or she then runs to the front of the group, and repeats the process until a given time is exhausted.

The group to transport the largest volume of water wins!

### Possible Variations

A possible variation of the game could be to place two buckets side by side. One of the buckets is filled with water. Players are not to step within 1.5m radius of the buckets.

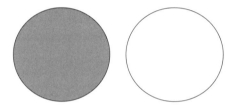

Players are to make use of all the logistics given to create a method to transfer water from the filled bucket to the empty bucket within a given time. If less than half of the empty bucket is filled, players will have to do a forfeit as it implies that their "invention" or method of transporting the water is not efficient enough.

### Process

- If you were to do this again, what would you do differently?
- What did you do to ensure that you spilled as little water as possible?
- How did you feel about the amount of water in the test tube when it was first handed to you?

*Practical Application*

- If the test tube represents your predecessor's work, how do you feel about work left over for you to take over?
- Was the bucket half full or half empty? How will others perceive the work that you leave behind?
- What is succession in leadership? What role can you play to ensure good succession?

### Game 41: Water Bomb Volleyball

1. Each team will have their group members paired up, with a garbage bag between each pair.

garbage bag

2. Use the garbage bag to fling the water bombs over the net to the opponent's side.

3. The opponents try to receive the bomb using the garbage bag. If they fail to do so, the attacking team gets 1 point.

Team with most number of points wins.

# Leadership Game 41: Water Bomb Volleyball

*Key Leadership Understanding*

Leaders observe changes to circumstances and adapt well.

*Math/Science Concepts Applicable*

Estimation of forces

*Equipment/Logistics*

Garbage bags
Towels
Water bombs
Net

*Time Required*

30 minutes

*Game Objective*

Be the group to score as many points as possible while preventing the other team from scoring

*Group Size*

Eight

*Procedure*

Each team is separated by a net. Each team will have their group members paired up, with a garbage bag between each pair. The objective of the game is to use the garbage bag to fling the water bombs over the net to the opponent's side.

If the water bomb falls to the ground and bursts, the team that throws the water bomb gains a point. If the opponent pair manages to catch the water bomb with their garbage bag, the opponent pair gains a point. Within each team, the water bomb can be passed around.

The team with the highest points in the given time wins!

### Possible Variations

Other items other than water bombs can be used, such as a beach ball, etc.

### Process

- How did you feel when the water bomb burst?
- What was essential for this activity to be a success?
- Did you have problems communicating with your team members so that the roles can be evenly distributed out? (e.g., the positioning of pairs, etc.)
- Given a chance to repeat the activity, what would you have done differently?

### Practical Application

- What does this activity tell us about a leader's role in such a circumstance?

**Game 42: Swoosh!**

Bird's eye view of chairs.

1. Person from each team tries to slide across chairs as fast as possible.

2. Facilitator gives them a science question when they meet.

3. Try to answer as fast as possible. Winner gets to continue sliding, loser sends out another person.

4. Team with group member that reaches the end of the bench wins!

# Leadership Game 42: Swoosh!

*Key Leadership Understanding*

Leaders set the pace for everyone and keep the momentum going in the team.

Leaders encourage one another.

*Math/Science Concepts Applicable*

Forces of friction

*Time Required*

15 to 20 minutes

*Equipment/Logistics*

Three benches
Soap water
One pail
List of science questions

*Game Objective*

Be the group to score as many points as possible by answering the science questions in the shortest time

*Group Size*

20

*Procedure*

It would be good to carry out this activity in an outdoor court. Benches will be set up in the format below:

Benches

Team A

Team B

Soap water will be splashed on the benches to make the surface slippery.

Each team will have to send out one member at a time to slide themselves on the benches as fast as possible. The players will start in opposite directions. At the point when two members of the different teams meet, the facilitator will ask the two members a science question. The first player to answer the question correctly is allowed to continue sliding on the bench heading towards the opposing team. The player who could not answer the question fast enough will then have to come off the bench immediately so that another member of his team can start sliding to meet and prevent the "winner" from going further.

The aim is to reach the end of the benches at your opponent's end to score a point for your team.

The team that has the most number of players at the other end wins!

*Process*
- Why did you think you won or lost the game?
- Was there any strategizing involved in this game?
- Did you play the game as fairly as you could?

*Practical Application*
- What are the strengths and weaknesses of your team?
- What are the ways we can encourage others to help them reach their goals?

### Game 43: Leucocytes

Try to form a cell with its parts (e.g. nucleus, cytoplasm, etc) as well as a "virus". Try to engulf the virus of another group while protecting your own virus.

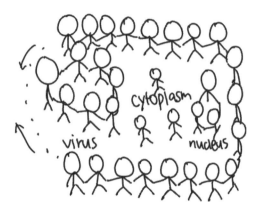

White blood cell engulfing virus

# Leadership Game 43: Leucocytes

*Key Leadership Understanding*

Leaders care for his or her team members throughout the journey.

Leaders embrace diversity.

*Math/Science Concepts Applicable*

Cells and viruses

*Time Required*

20 minutes

*Equipment/Logistics*

A big open space with a marked out boundary

*Game Objective*

Be the group that successfully protects their own virus

*Group Size*

20 to 30

*Procedure*

Divide group of players into Teams A and B. All the players must form a cell with its parts (a few players to represent the virus, cytoplasm, cell membrane, white blood cell or the nucleus). Both teams will need to strategise their plan to engulf the "virus" members of the opposing team.

There should at least four "virus" members and the number of such members must be equal in each team.

All the players will be stationed in the marked out boundary area. Team A and Team B will each offer and place a "virus" member in the centre of the boundary area. When the facilitator starts the game, teams will rush to the centre to attempt to capture the opponent "virus" member while protecting their own "virus" member from being captured.

This game will continue until only one "virus" is left. The team with the remaining "virus" wins!

### Process

- What was the most essential factor that allowed you to complete the game?
- How did this game show you the importance of communication and cooperation with other team members?

### Practical Application

- What kind of situations calls for encouragement to be given and how can we best provide it?
- Was it easy to embrace your opponent's "virus"? How could we improve our ability to embrace differences in others?

## Game 44: Charades

1. 1 representative will be shown a scientific term.

2. That person must act out the word.

Huh?

3. He can get another to act it out. but he cannot guess the word.

??? 

4. If the word is guessed correctly another term would be given.

Yay!

# Leadership Game 44: Charades

### Key Leadership Understanding

Leaders communicate well and think out of the box. Leaders persevere in the face of challenges.

### Math/Science Concepts Applicable

Knowledge of scientific terms

### Equipment/Logistics

List of scientific terms (e.g., Hall effect, etc.)

### Time Required

Five minutes per team (total time is dependent on the number of teams playing)

### Game Objective

Be the group to guess correctly as many terms as possible within a stipulated time (e.g., one minute)

### Group Size

Five

### Procedure

Each team will send out one representative. The representative will be shown a scientific term. He or she has to act out the term for team members to guess what the term is. The representative is not allowed to talk or write. Once the word is guessed correctly by team members, the representative will then be given another term to act out.

If the representative is unsure of how to act out the term, he or she can choose to pass it to another team member to replace him. However, he or she is not eligible to guess that word anymore. Alternatively, the representative can choose to skip a term and ask for another term to act it out.

## Process

- How did you (the person acting out the words) feel when you were unable to transmit the message effectively to your teammates?
- How did you (the rest of the team) feel when you were unable to guess the terms?
- Were there other ways which you could have helped the transmitter? (e.g., asking questions, etc.)

## Practical Application

- When do boldness and honesty in communication create problems for the team?

## Game 45: Drinks for Life

70% orange juice?
60%?
50%?

# Leadership Game 45: Drinks for Life

*Key Leadership Understandings*

Leadership involves prioritising.

Leaders understand that their ideas and choices affect others.

*Math/Science Concepts Applicable*

Observation and deduction

*Equipment/Logistics*

Four types of liquids – cola with phosphoric acid, vinegar with acetic acid, sodium chloride with salt and sodium carbonate with baking soda

*Time Required*

30 minutes

*Game Objective*

Be able to guess the composition of liquids used in a mixture

*Group Size*

Four to six

*Procedure*

In groups of three, players will be given four liquids. Each group will select any two of the liquids to create a unique mixture.

Before they create their mixture, they will have to guess what will result from their mixing. Will there be a precipitate? They would need to state the colour and taste (e.g., sour, sweet, acidic, etc.) of their mixture.

After each group has completed making a guess, they will proceed with the mixing. The group will then compare the outcome of the mixture with what they have inferred prior to the mixing.

The facilitator will allocate some time for each group to share how they came to their inferences and how did the outcome differed from their inferences.

*Process*

- How did you feel when you were making the unique drink?
- Was it easy or difficult to make the selection of liquids to be used in the mixture? Why or why not?
- What happened to the original liquids after they were mixed?

*Practical Application*

- What guides you in making choices?
- Why is it difficult to make a choice at times?
- What effect do our decisions make on the lives of others?

### Game 46: Treasure Hunt

1. Decipher the coordinates given based on the list of clues given to you.

2. Find the cards hidden at that particular place as described by the coordinate, buried in the sand.

"coordinate (4, 1)!"

3. Each card has a letter. Piece it to solve the motto together.

# Leadership Game 46: Treasure Hunt

### Key Leadership Understanding

The true treasures lie in the qualities of each member in the team. Leaders encourage and motivate others in search of such treasures or qualities.

### Math/Science Concepts Applicable

Science based concepts

### Equipment/Logistics

Pieces of cards, one with each letter (all the letters will eventually form a word that highlights a particular theme or concept, e.g., the letters *u, n, i, t, y* for *unity* in leadership)

Sets of coordinates

List of scientific questions for each group to solve in order to have the coordinates

Sandpit where the cards can be buried

One game master for each group to assign the numerals for each correct answer

### Time Required

More than an hour (depending on the difficulty level of the questions)

### Game Objective

Be the first group to successfully piece the letters to form the theme

### Group Size

Six to eight

## Procedure

On the sandpit, facilitator will mark out the coordinates and bury the pieces of paper with the letters in the sand in each of the coordinates to be deciphered.

Provide each group with the same list of questions where each answer points to a single number. The groups will need to answer the questions in order to find the numerals, and piece them together to form sets of coordinates.

For instance, the theme "United in Leadership" involves 18 letters all of which will be buried in the sandpit. The entire list of 36 questions is shared equally among the number of groups. Their correct answer to any one question will yield them one numeral. The correct answer to another question will yield them another numeral. These two numerals will form one set of coordinates. Each set of coordinates will lead them to one letter in the sandpit.

NOTE
Every group will receive a unique set of coordinates with each pair of correct answers. The game master for each group will assign the pre-determined sets of coordinates which can be found on the sandpit.

Each group will have the same number of letters which are insufficient to form the word or theme. They will then have to communicate with other groups to pool their letters together.

## Possible Variations

Number of letters can be reduced if there is not enough time. Have a shorter form of the theme. Alternatively, the difficulty level of the questions can be made easier.

## Process

- Where is the "treasure" in the hunt?
- What challenges did you face in this activity?

- If you were to do it again, what would you do differently?
- Who was the most encouraging player in the team?

*Practical Application*

- Which are the difficult situations in your organisation where it might be easy to get discouraged?
- How can we keep people moving forward in changing for the better?
- What conditions are necessary for high performance in a team?

## Game 47: Table of Elements

Revise the periodic table to reflect key leadership terms.

"Humility..."

# Leadership Game 47: Table of Elements

*Key Leadership Understanding*

Leaders resolve problems by looking at and perceiving things differently.

*Math/Science Concepts Applicable*

Natural elements

*Equipment/Logistics*

Periodic Table of Elements

*Time Required*

30 minutes

*Game Objective*

Be the first team to revise the Periodic Table of Elements to reflect key leadership terms

*Group Size*

Three to five

*Procedure*

Give group of three participants the Periodic Table of Elements. Participants have to revise the Table to reflect new leadership terms or principles.

Example: *Fe* for *Iron* becomes *Fe* for *Fear*

## Process

- How did you feel trying to revise the terms?
- What challenges did you face in revising the terms?
- How would you do it differently the next time?

## Practical Application

- What does this challenge relate to the difficulties of leaders in managing team effort?
- What are the difficult situations you are currently facing, where it might be easy to get discouraged?
- Which qualities are necessary to press on?

## Game 48: Primordial Soup

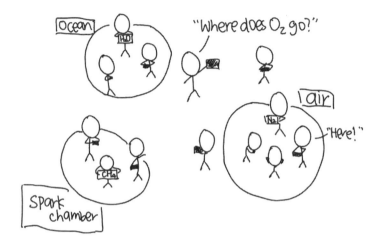

# Leadership Game 48: Primordial Soup

*Key Leadership Understanding*

Leaders aim for and set high standards.

*Math/Science Concepts Applicable*

Chemical compositions

*Time Required*

30 minutes

*Equipment/Logistics*

Set of placards A with a chemical symbol (e.g., $H_2O$ on one placard, $O_2$ on another, $CH_4$, C, $N_2$, NaCl, rare gases) on each placard

Set of placards B indicating a percentage (e.g., 7%, 21%, 0.03%, 3.4%, 79%, 96.6%) on each placard (each percentage represents an actual chemical composition)

Set of placards C indicating another chemical symbol (e.g. amino acids, etc.) to serve as distractors

Twine/raffia string (to mark out circles)

*Game Objective*

Be the first group to come up with the correct chemical composition of the ocean, spark chamber and air

*Group Size*

Eight to ten

*Procedure*

Similar to the memory game, all the placards are to be placed face down so that players will have to flip the placards over to view the chemical (A), percentage (B) or distractor (C) symbols. Physically mark out three circles with space big enough for five to six players (one circle to represent the ocean entity, one to represent the spark chamber entity and one to represent the air entity). All placards are to be scattered outside of the three circles.

One member of the team starts the game by picking up a placard, and decides which circle to go into with the placard. One team member is only allowed to enter into the circle at any one point, and can only flip over the placard in the circle. If the symbol on the placard does not fit the circle, the team member must leave the circle and return the placard to its original place. The next team member follows this procedure.

Eventually, the teams would have correctly placed all the placards in the respective circles, with the chemical composition of either the ocean, air or spark chamber.

The facilitator has the answer sheet of each chemical symbol in each of the circle. The fastest team to correctly have the right chemical composition for all the three circles wins. Answers should be scientifically correct.

NOTE
Air = 79% $N_2$, 21% $O_2$, 0.03% $CO_2$ and rare gases
Ocean = 96.6% $H_2O$ and 3.4% NaCl
Spark Chamber = $O_2$ and $CH_4$

*Process*

- Did you have difficulty knowing where each chemical was supposed to be in?
- How did those who know the information communicate to those who do not?
- Did you feel impatient with any group member who might have gone into the wrong circle?

*Practical Application*

- Are we impatient for results?
- Are we expecting too much from others to attain our high standards?
- Do we feel competent enough to take on a leadership role?
- How do we prepare ourselves as leaders of our team or group?

### Game 49: Height Equilibrium

1. Start off with 1 person in the arena.

2. Put in another person in the arena.

3. The game master will announce a particular height that the players have to attain either by elevating or lowering themselves.

4. Repeat steps 2 and 3:

"I want all of you to be as tall as him."

# Leadership Game 49: Height Equilibrium

*Key Leadership Understanding*

Leaders are resourceful and adapt well.

*Math/Science Concepts Applicable*

Heat transfer generally occurs from a hotter to a cooler object until both objects are at the same temperature. This state is deemed as thermal equilibrium. In this game, all the players are supposed to try and reach a state of "height equilibrium".

*Time Required*

However long it takes to achieve height equilibrium

*Equipment/Logistics*

Any physical object (e.g., a chair or table) which could be used as a benchmark for the height required by the facilitator

*Game Objective*

Be the fastest group to achieve the height equilibrium

*Group Size*

Six to eight

*Procedure*

The game starts with only one player from each group assigned to an arena. The Height Almighty (HA) or facilitator will provide a particular height for the player to attain. The player could either squat, sit, stand or use the available logistics (e.g., a chair or table) given. The team is allowed to introduce the next player only when the HA is satisfied with the height achieved by the player(s) in the arena.

The fastest group to attain "height equilibrium" with all members in the arena wins.

*Process*

- When some of your teammates were not tall enough, how did you overcome the problem? Do you think it was effective?
- Was there a strategy that you used in this activity?
- What were some of your feelings when you were unable to attain a certain height?
- Do you feel that you have encouraged each other enough?

*Practical Application*

- What kind of members do we want to recruit for our groups or teams to make things stronger or better?
- What plans could we make to achieve what is best for our team?

**Game 50: Pull It Up!**

1. Groups are given a list of items that they have to transport up to the 2nd storey.

"Who has pink earrings?"

2. Using whatever they have on themselves, they build a pulley-system to transport the items up.

3. The 1st team that transports everything up wins!

# Leadership Game 50: Pull It Up!

*Key Leadership Understanding*

Leaders communicate through different ways to inspire others in achieving a shared vision.

*Math/Science Concepts Applicable*

Newton's Second Law in physics

*Equipment/Logistics*

Strings
Scissors
List of items consisting of things that can be found on players (e.g., shoes, socks, watches, etc.)

*Time Required*

About an hour dependent on the number of items

*Game Objective*

Be the fastest group to build a pulley, transfer items according to instructions

*Group Size*

10 to 12 (five to six members on each level)

*Procedure*

Each group will be asked to provide a list of items from their belongings. They are given strings, scissors and whatever they have to build a pulley system. Each team has to split into two, with one group stationed on the ground level and the other on the second level of a building.

The group at the ground level will be given the items which they have to transfer using the pulley system to their team members on the second level.

After all the items have been transferred, the pulley system has to be dismantled and all the members will assemble at the ground level to complete the game.

The fastest group to complete wins.

## Possible Variations

Certain communication restrictions, such as no shouting, can be imposed on the team.

## Process

- Was it hard building the pulley system? Why or why not?
- Was communication tough? Could there be other means of communication apart from shouting (if shouting was used)?
- How could you do this better the next time round?

## Practical Application

- How do we involve reluctant members in meeting the team's goals?